story telling MATH

Lia & Luís

Puzzled!

Ana Crespo

Illustrated by
Giovana Medeiros

Charlesbridge

Lia and Luís receive a package from their grandma. Inside there's a puzzle . . .
Oba!
A

. . . with a secret message.
Lia loves a mystery.
Luís does not.

Still, he's curious
about Vovó's gift.

But Lia and Luís must hurry.
Mamãe says they're leaving soon.
Where are we going?
A

You'll find out.
Just finish the puzzle!

Lia takes the lead.
She inspects the pieces closely.

They're all different shapes.

But some pieces have a straight side.

It's a clue!

Luís helps Lia sort the pieces.
He finds another clue.

There's a piece with *two* straight sides.

So they start another pile.

It's time to reveal Vovó's secret message. They begin connecting pieces with just one straight side.

But it's not that simple.

Luís is about to lose his patience . . .

. . . when Mamãe makes them take a break.

In the hurry to get ready . . .

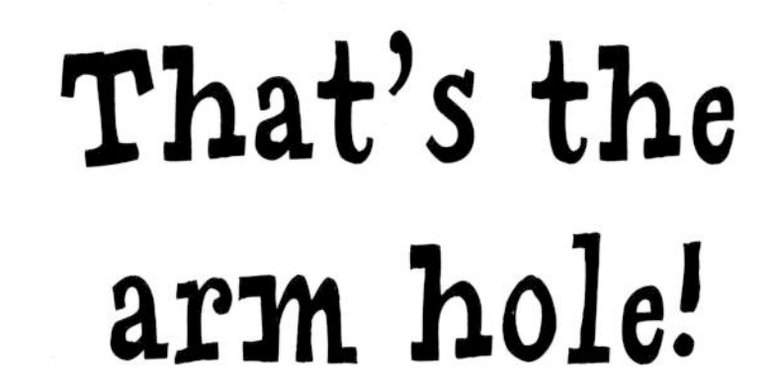

. . . they make a breakthrough.

For the pieces to connect, each part must be just the right size . . .

. . . just the right shape . . .

. . . and just the right color.

Finally, Lia finds just the right piece.

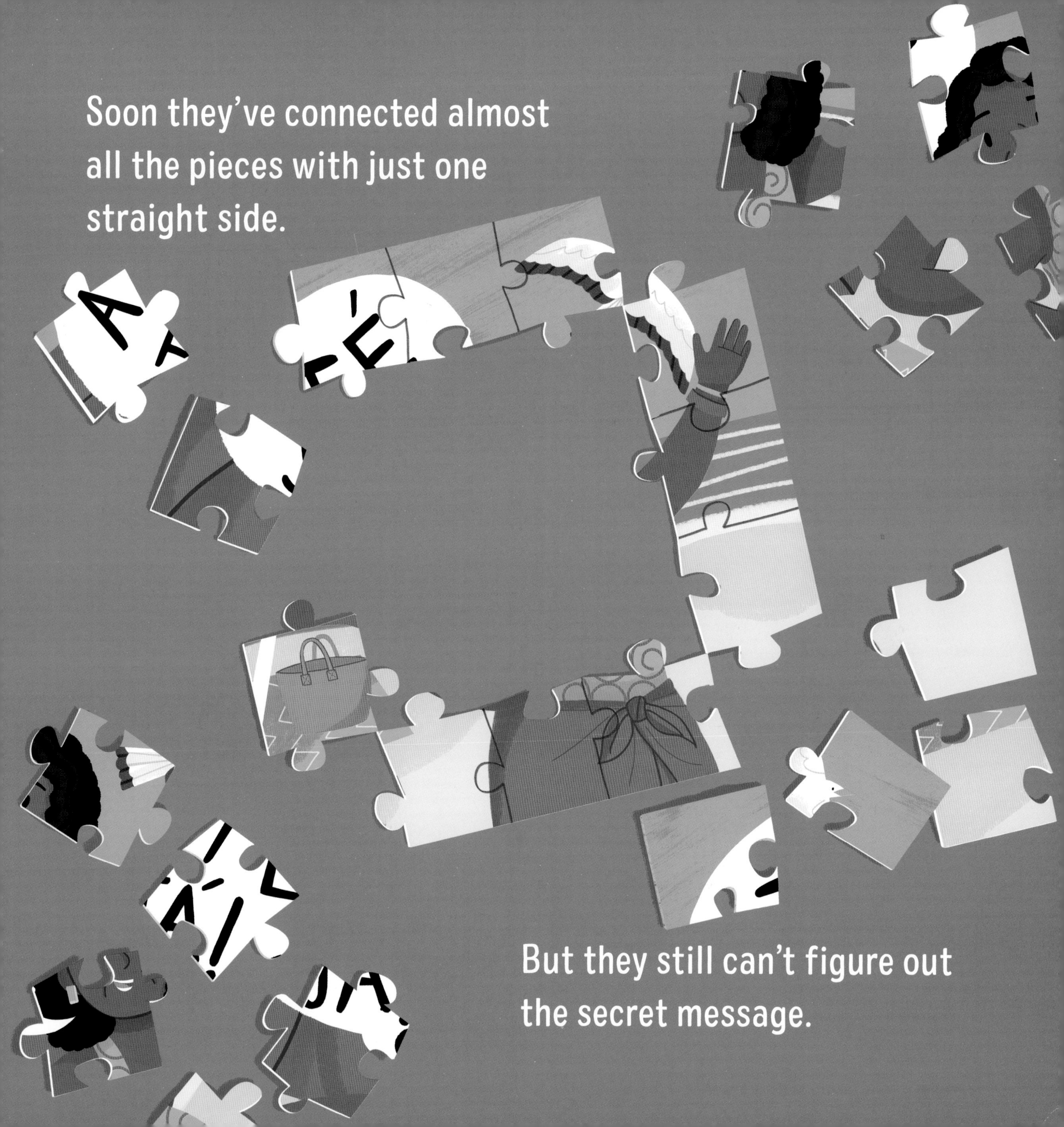

Soon they've connected almost all the pieces with just one straight side.

But they still can't figure out the secret message.

Luís has an idea. He tries the pieces with two straight sides. There are four of them.
They're corners!

The edges are finished.
Five minutes!
Now they must focus on the middle.

Sometimes the shapes seem to fit, but the colors don't match.

Sometimes the colors seem to match, but the shapes don't fit.

And sometimes . . .

they must turn . . .

and turn . . .

and turn the same piece . . .

until it fits perfectly.

ATÉ JÁ!

Vovó is coming
to visit!

Lia and Luís have solved the mystery.

It's time to pick up Vovó.

Que saudade!
We missed you too, Vovó!

GLOSSARY

People all over Brazil speak Portuguese, but they have different local accents. Lia and Luís's Brazilian American family speaks with the accent of people from São Paulo.

AUTHOR'S NOTE

It's not uncommon for immigrant children or the children of immigrants, like Lia and Luís, to live far away from their grandparents. Since I immigrated to the United States, technology has advanced, allowing my family and me to see one another through computer screens. But there's nothing like hugging a grandparent after years away and ending (at least for a moment) the saudade we all feel.

EXPLORING THE MATH

As Lia and Luís piece together a jigsaw puzzle to reveal its secret message, they explore the math of spatial relationships. They compare characteristics of puzzle pieces, such as straight sides and bumpy knobs. They consider how various shapes and sizes fit together and rotate puzzle pieces in different ways. They also use strategies like sorting the pieces and assembling the edges of the puzzle first.

Assembling jigsaw puzzles is a wonderful way to develop spatial skills, which are important in math, science, and the visual arts.

Try this!

- **Make your own puzzle.** Cut a drawing, printout of a photo, or side of a cereal box into pieces for children to put back together.
- **Talk about what fits.** As children put together the puzzle, invite them to explain how they know which pieces fit together and which don't.
- **Choose a secret puzzle piece.** Encourage children to try to identify it by asking yes-or-no questions, such as "Does it have any straight sides?"
- **Play "I Spy."** One person gives a clue about color and shape: "I spy something round and black." The others find things that match the clues.

As you go about the day, look for opportunities to help children notice similarities and differences among shapes. Observing and describing objects encourages children to think like mathematicians, scientists, and artists!

—Ximena Dominguez, PhD
Director of Early STEM Research, Digital Promise

Visit **www.charlesbridge.com/storytellingmath** for more activities.

To Alyssa and Marlene: Thanks for working on this puzzle with me.—A. C.

To "Vó Otília," who always inspires me so much.—G. M.

This book is supported in part by TERC under a grant from the Heising-Simons Foundation.

Developed in conjunction with TERC
2067 Massachusetts Avenue
Cambridge, MA 02140
(617) 873-9600
www.terc.edu

Published by Charlesbridge
9 Galen Street
Watertown, MA 02472
(617) 926-0329
www.charlesbridge.com

Printed in China
(hc) 10 9 8 7 6 5 4 3 2 1
(pb) 10 9 8 7 6 5 4 3 2 1

Library of Congress Cataloging-in-Publication Data
Names: Crespo, Ana, author. | Medeiros, Giovana, illustrator.
Title: Lia & Luís: puzzled! / by Ana Crespo; Illustrated by Giovana Medeiros.
Description: Watertown, MA: Charlesbridge, [2023] | Series: Storytelling Math | Audience: Ages 3–6 | Audience: Grades K–1 | Summary: "When Brazilian American twins Lia and Luís receive a jigsaw puzzle from their grandmother, they must quickly solve it to figure out its secret message."—Provided by publisher.
Identifiers: LCCN 2021031421 (print) | LCCN 2021031422 (ebook) | ISBN 9781623543228 (hardcover) | ISBN 9781623543235 (paperback) | ISBN 9781632899804 (ebook)
Subjects: LCSH: Problem solving—Juvenile literature. | Mathematical recreations—Juvenile literature.
Classification: LCC QA63 .C74 2023 (print) | LCC QA63 (ebook) | DDC 793.74—dc23
LC record available at https://lccn.loc.gov/2021031421
LC ebook record available at https://lccn.loc.gov/2021031422

Illustrations done in digital media
Display type set in Strike One by Creativeqube and Handegypt by Matt Desmond
Text type set in Colby Narrow by Jason Vandenberg and Handegypt by Matt Desmond
Printed by 1010 Printing International Limited in Huizhou, Guangdong, China
Production supervision by Jennifer Most Delaney
Designed by Jon Simeon